FINITE THEORY

HISTORICAL MILESTONE IN PHYSICS

PHIL BOUCHARD

Copyright © 2020 by Phil Bouchard

All rights reserved. This book or any portion thereof may not be reproduced or transmitted in any form or manner, electronic or mechanical, including photocopying, recording, or by any information storage or retrieval system, without the express written permission of the copyright owner except for the use of brief quotations in a book review or other noncommercial uses permitted by copyright law.

Printed in the United States of America

Library of Congress Control Number:	2019919413
ISBN: Softcover	978-1-64376-647-8
eBook	978-1-64376-646-1

Republished by: PageTurner, Press and Media LLC
Publication Date: 02/07/2020

To order copies of this book, contact:

PageTurner, Press and Media
Phone: 1-888-447-9651
order@pageturner.us
www.pageturner.us

Contents

1 Preface vii
2 Acknowledgments ix
3 Abstract xi
4 Special Relativity 1
    4.1 Lorentz time transformations 1
    4.2 Twin paradox 2
    4.3 Length contraction paradox 3
    4.4 Side by side comparison 4
5 Foundation of the Finite Theory 7
    5.1 Hypotheses of the Finite Theory 7
    5.2 Black hole radius 8
    5.3 Gravitational time contraction 8
    5.4 Universe of 1 galaxy 14
    5.5 Universe of 2 galaxies 17
    5.6 Dynamic universe of 2 galaxies 20
    5.7 Time dilation effect 23
    5.8 Bending of light in the gravitational field 25
    5.9 Explanation of the perihelion shift 26
    5.10 GPS and time dilation cancellation altitude 27
    5.11 Gravitational redshift 29
6 Cosmological implications 33
    6.1 Natural faster-than-light speed 33
    6.2 Parameters of the invisible universe 35
    6.3 Galactic rotation curve 36
    6.4 Approximation of the center and velocity of the visible universe 38
7 Experiment proposal 41
8 Conclusion 43
References 45

- 1 -
Preface

This book is an alternate proposal on the mathematical model already suggested by General Relativity by Albert Einstein this December 1916. It should fit to any the reader detaining a minimum of a bachelor degree in either science or engineering including a basic Newtonian understanding of the object dynamics, with roots in astrophysics and knowledge of the problems which are still unsolved to this day that are basically subject to many science fiction books and movies.

The author is assuming the formulation is presented as clear, concise but mostly as simple as possible and hopes to bring a different approach towards the subject by rejecting previous hypothetical effects leaked from General Relativity and by proposing a model easier to understand than its predecessor.

The text is very minimal and therefore makes the overall book faster to read. Most of the equations can easily be visualized by an adjacent graphical representation of the plot which should speed up greatly its understanding. Sketches were also added to explain potential conundrum in distinct scenarios.

We wish here the best for all researchers adventuring into this subject and luckily the theory will help linking astrophysics with quantum mechanics by unifying all forces.

December 2008
Phil Bouchard

- 2 -
Acknowledgments

This article was still a pure theoretical facet in the year 2005 and was fostered into a serious labor after confirmation of its potential validity by Dr. Griest and important requirements. Many thanks to him.

Furthermore my father, Mr. Bouchard M.Sc. Physics, brought considerable help in asserting the mechanical part of my equations. He also introduced me to the astrophysical society for advanced research.

The same goes directly and indirectly for the online scientific community, where we can find Cosmoquest.org and InternationalSkeptics.com, in order to have the opportunity to break the ice and debate theories that are against the mainstream.

Finally thanks to Dimtcho Dimov some help solving an equation and Evan Adams for helping editing it and for reviewing related experiments.

- 3 -
Abstract

The mathematical representation of General Relativity uses a four dimensional reference frame to position in time and space an object and tells us time is a linear variable that can have both a negative and positive value. This therefore implies time becomes itself a dimension and causes the theory opening doors to ideas such as: singularity, wormhole, paradoxes and so on.

In this book a new mathematical model is being suggested which is based on the classical mechanics. The theory is objective and predicts low scale GPS gravitational time dilation, perihelion precession disparity, gravitational light bending, artificial faster-than-light motion, up to the rotation curve for all galaxies, natural faster-than-light galactic expansion and can consequently be used to determine the ultimate scale of the universe.

In practical terms, this means:

- We can achieve infinite speed using basic nuclear fusion reactors with the adequate technology to cancel the effects of the gravitons.
- We can also attain time travel into the future with some small pod having an increased flux of gravitons traveling through.
- It follows levitation also derives from the theory.

A wavelength meter in motion is proposed to test directly the invariance of c as postulated by the Special Relativity, which is the first time this experiment is attempted [2]. Until now it was assumed aether, if it was found, was a static substance having a unique reference frame from which entities were traveling through and therefore must not be present if tests proved otherwise. If we replace aether with graviton fields overlapping each other then we will have a reference frame that follows the rotation of the Earth. Thus to detect its presence, we will have to physically move against that rotating frame in order to detect a change in speed of light. This is done by sending a laser beam in the same direction of the velocity vector of the moving apparatus, capturing the difference in wavelength as we will demonstrate in this proposal.

- 4 -
Special Relativity

You will find here the necessary mathematical arsenal currently available for estimating some effects we should expect. We are more interested into time dilation effects than length contractions because it is a measurable and thus tangible.

The solutions presented in this chapter do not use the complete General Relativity modeling because of its unnecessary complexity. What is being used is the simplified Special Relativity formulation that is basically the roots of its successor and therefore should be necessary to present the problem and the approach that should be taken to resolve it.

Quick presentations are made and it is assumed knowledge of their origin is known.

4.1 Lorentz time transformations

The accepted and most precise equation explaining time dilation is given by a derivation of the famous Lorentz transformations:

$$t_f = \frac{t_o}{\sqrt{1 - \frac{v^2}{c^2}}} \qquad (1)$$

Where:
- v is the relative velocity between the observer and the moving clock
- c is the speed of light

This equation is represented in figure 1.

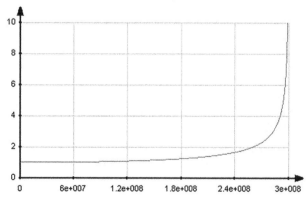

Figure 1: Time Dilation Factor vs. Speed (m/s)

4.2 Twin paradox

Special Relativity is a subjective approach to demonstrate effects caused by high-speed velocities and instantly leads to paradoxes such as the famous twin paradox. Taking time dilation into account when comparing accelerated frames altogether is fundamental.

By taking for example a variation of the twin paradox and consider the simplified scenario of a spaceship traveling nearly the speed of light in relation to stationary clocks:

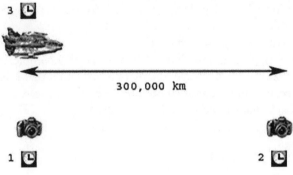

Figure 2: Spaceship traveling 300,000 km in 1 second

According to Special Relativity if clock 1 and 2 are perfectly synchronized and their adjacent camera takes a picture of the spaceship at $t = 0$ and $t = 1$ then they will both return the picture of a spaceship having clocks indicating a time almost identical.

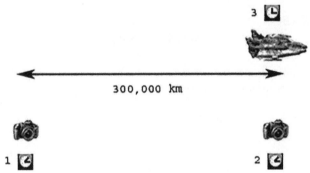

Figure 3: Observed time dilation after 1 second

Where Special Relativity fails is to represent the perception of our spaceship. Pursuant to Special Relativity:

"The laws by which the states of physical systems undergo change are not affected, whether these changes of state be referred to the one or the other of two systems in uniform translatory motion relative to each other."

Well because of the time dilation effects we can see the lack of change in time on the clock number 3. This means the observer driving the ship *must* see the external clock 1 and clock 2 running very fast in relation to his own. Thus the time outside his reference frame will be viewed as being nearly infinitely much quicker.

4.3 Length contraction paradox

Having a paradox in the heart of a theory ultimately disproves it. Notwithstanding infinite masses is also predicted by SR, length contraction is enough showing inconsistency.

According to SR length of a body will contract in the direction of its velocity when its amplitude approaches the speed of light as we can see from the formula:

$$L' = L\sqrt{1 - \frac{v^2}{c^2}} \qquad (2)$$

If we study further in details the behavior of this effect by having two cannons adjacent to each other and an observer taking in picture the length of the objects thrown by the cannons, then we will get very different results out of distinct scenarios.

In essence a projectile when traveling at high velocity as predicted by SR will have a velocity of $c - \varepsilon \; m/s$ in our example:

Figure 4: Length contraction affecting single projectile

Thought experiment #1

We then decide adding a second cannon next to the first one and exactly 1 meter in front of it. A clerk behind the cannons now fires the cannonballs at the exact same moment and the photographer standing on the ground should measure the distance of the two bullets to be 1 *meter*.

Figure 5: Cannons 1 meter away, simultaneous firing

By now tying the bullets with a chain, the photographer won't see any difference in the distance of the two bodies in motion:

Figure 6: Cannons 1 meter away, simultaneous firing of tied projectiles

Thought experiment #2

The leading cannon in this scenario is moved back and placed side by side to the other one. 3.33X10^{-9} *second* after the first cannon fires, the second one is then ejected making it 1 *meter* away from the leading one:

Figure 7: Cannons juxtaposed, asynchronous firing

Up to now everything is consistent with SR but here comes the tricky part. What would exactly happen if the two bullets were tied together with a rope before they are ejected from their respective cannon as shown below?

Figure 8: Cannons juxtaposed, asynchronous firing of tied projectiles

This is controversial because now the 2 bullets become 1 object only. According to SR this basically means the 2 bullets and the rope must all contract altogether.

It makes the two last events incongruous by defining the behavior of the objects already in motion with any precondition.

4.4 Side by side comparison

Given from the gravitational time dilation of General Relativity:

$$t_o = \frac{\sqrt{1-\frac{2Gm}{|i|c^2}}}{\sqrt{1-\frac{2Gm}{(x+|i|)c^2}}} \times t_f \qquad (3)$$

And from the gravitational time dilation of Finite Theory as later stipulated:

$$t_o = \frac{\frac{m}{|x-i|}+h}{\frac{m}{|i|}+h} \times t_f \qquad (4)$$

If we equate the aforementionned equations by using a reference point infinitely far away and letting h include the effects of the entire universe:

$$\frac{\left(\frac{m}{r}+h\right)^{-1}}{h} = \frac{\sqrt{1-\frac{2Gm}{rc^2}}}{1} \qquad (5)$$

$$\left(1+\frac{m}{rh}\right)^{-1} \approx 1 - \frac{Gm}{rc^2} \qquad (6)$$

$$\left(1+\frac{m}{rh}\right)^{-2} = 1 - \frac{2Gm}{rc^2} \qquad (7)$$

We will observe that General Relativity is making use of a constant in its equations:

$$h = \frac{c^2}{G} = 1.35 \times 10^{27} \; kg/m \qquad (8)$$

As it turns out, this happens to follow the same ratio used by Einstein's constant:

$$\kappa = \frac{8\pi G}{c^2} \qquad (9)$$

- 5 -
Foundation of the Finite Theory

5.1 Hypotheses of the Finite Theory

Finite Theory defines a new representation of the formulas derived from General Relativity based on the superposed potentials of the predicted massless spin-2 gravitons that mediate gravitational fields.

Additionally in contrast to General Relativity where the space-time is represented using the non-Euclidean geometry in order to keep the speed of light constant, Finite Theory considers time to be a positive variable within a space that is characterized by the Euclidean geometry.

Hypotheses of the Finite Theory are as follows:

Definition 1: A 'comoving framework' moves coherently with the source of the strongest gravitational acceleration amplitude. For example, if the observer and the observed object are nearby a planet then the comoving framework is set on the planet's center, rotating with the same angular speed. Note that this can be a non-inertial frame.

Definition 2: A 'parent framework' is the source of the 2nd strongest gravitational acceleration amplitude. The source here is a collective noun and represents the conglomeration of its constituents.

Definition 3: An 'absolute framework' is a comoving framework that has no parent framework.

Definition 4: The kinetic energy is defined as $1/2mv^2$ (classical definition), with being the speed of the object with respect to the observer.

Definition 5: A gravitational time dilation is directly proportional to the ratio of the superposed gravitational potentials of the observer and the observed object.

Hypothesis 1: The speed of light in free space has value c for any observer at rest relative to the comoving framework. However, observers in relative motion with respect to this frame will not measure the same value for c.

Hypothesis 2: The time dilation experienced by an object moving with respect to an observer at rest relative to the comoving framework is directly proportional to the ratio between the kinetic energy and the maximum kinetic energy of the object, where the latter is the case when its speed equals c.

Below, we'll consider consequences from these hypotheses to the time dilation effect.

5.2 Black hole radius

A black hole is a region in space where all matter and energies, including light, cannot escape from its gravitational force. The Schwarzschild radius defines the event-horizon where the gravitational pull exceeds the escape velocity of the speed of light. This is given by:

$$r_s = \frac{2GM}{c^2} \qquad (10)$$

Given that Schwarzschild radius derives from GR formulation, FT will need its own definition. Satisfyingly, this event horizon can easily be found with the amount of kinetic energy needed to overtake the gravitational potential energy:

$$\lim_{v \to c} \frac{1}{2}mv^2 = \frac{GMm}{r_b} \qquad (11)$$

By solving the equation with the maximum escape velocity a photon can have, where the mass is of non-importance we get:

$$r_b = \frac{2GM}{c^2} \qquad (12)$$

5.3 Gravitational time contraction

Since in the candidate theory the acceleration is defined by gravitons pulling the body in the opposite direction of their velocity, the net effect of the gravitational acceleration already defines the flux. Unlike kinetic time dilation this is not an incident event but the residuum of the modus operandi by the acceleration vector magnitude.

In contrast to kinetic time dilation, gravitational time contraction will be used interdependently with the non-trivial ambient gravity field of the observer, or fractionalized.

Inverse square law – Method 1

Given that FT gravitational time dilation and the Newtonian gravity force are similar, the standard model of gravity inside a sphere cannot be directly linked with FT because factors applied in one direction will not cancel their equivalent in the opposite direction. This means no simplification can be made and all infinitesimal elements of the mass will affect the net amplitude at one particular location.

First, we can represent the respective factor with a triple integral in the following way, using spherical coordinates:

$$f = \int \int \sin(\theta) \int \frac{\rho^2}{r^2} d\rho d\theta d\varphi \qquad (13)$$

The square of the distance between the observer located at (x2, y2, z2) and the infinitesimal element being integrated is equivalent to:

$$r^2 = (r_1\sin(\theta)\cos(\varphi) - x_2)^2 + (r_1\sin(\theta)\sin(\varphi) - y_2)^2 + (r_1\cos(\theta) - z_2)^2 \quad (14)$$

By mapping the location of the observer from the center of the sphere on to an arbitrary axis with the same radius, we can simplify our denominator. In this case *(0, 0, d2)* will be used to map a radius into Cartesian coordinates:

$$f = \int \int \sin(\theta) \int \frac{\rho^2}{-2\rho * r_2 \cos(\theta) + r_2^2 + \rho^2} d\rho d\theta d\varphi \qquad (15)$$

$$f = \frac{\left(\pi \times \left((d_2^2 - r_1^2) \times \log((d_2 + r_1)^2) + (r_1^2 - d_2^2) \times \log((d_2 - r_1)^2) - 4r_1 d_2\right)\right)}{-2d_2} \qquad (16)$$

Where:
- *r1* is the spherical mass radius
- *d2* is the distance of the observer from the center

For example, the respective factor of an observer at position *x* inside a sphere of radius 20 *m* will be proportional to:

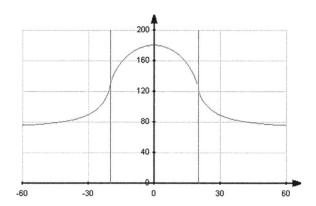

Figure 9: Inner Acceleration Factor vs. Radius (m)

Inverse square law – Method 2

Different means of calculating the inner gravitational time dilation factor with no relation with the aforementioned procedure can also be used. It consists of calculating the intersection between a growing sphere held within the spherical body in question.

This is done by first calculating all sphere surfaces fitting inside the largest sphere not in intersection with the spherical body. This represents the following area:

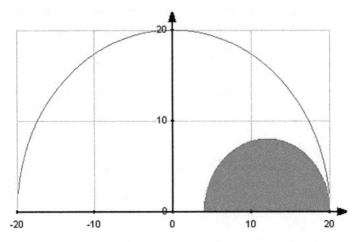

Figure 10: Inner Acceleration Factor vs. Radius (m) – Step 1

Now for the second part the spherical cap surface area resulting from the intersection of the two spheres will have to be considered only. This will cover the next section:

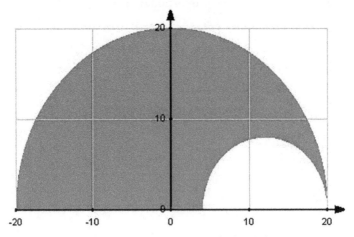

Figure 11: Inner Acceleration Factor vs. Radius (m) – Step 2

By summing both areas we will have:

$$f = \int \frac{2\pi r_2 \left(\frac{d_2^2 - r_2^2 + r_1^2}{2d_2} + r_2 - d_2\right)}{r_2^2} dr_2 + \int \frac{4\pi r_2^2}{r_2^2} dr_2 \quad (17)$$

$$f = \frac{\pi\left[d_2^2 \log\left(\frac{r_1+d_2}{r_1-d_2}\right) + r_1^2 \log\left(\frac{r_1-d_2}{r_1+d_2}\right) + 2d_2 r_1 - 4d_2^2\right]}{-d_2} + 4\pi(r_1 - d_2) \quad (18)$$

Where:
- $r1$ is the spherical mass radius
- $d2$ is the distance of the observer from the center

For instance, the same factor of an observer at position x inside a spherical body radius of 20 m will be corresponding to:

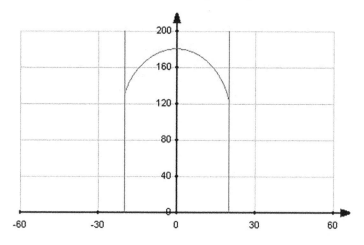

Figure 12: Inner Acceleration Factor vs. Radius (m)

This results in exactly the same inner curve as the one found by Figure 11. Henceforth this confirms the validity of the equation.

Juxtaposition

As a means to compare previous formulation with the more common Newtonian gravitational acceleration , we will before all else find the corresponding mass with the volume of a sphere with the respective radius at a given mass density ϱ. Hence:

$$m = \frac{4\pi r^3 \rho}{3} \quad (19)$$

The general acceleration formula contains a discrepancy constant we will call B:

$$a = \frac{Bm}{r^2} \qquad (20)$$

By solving the equation using the lowest factor found with Equation (18), the constant turns out to be 3/2 for a given mass density ϱ. By placing side by side both plots we have:

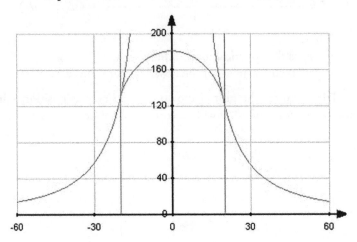

Figure 13: Inner & Outer Acceleration Factor vs. Radius (m)

Here, the outer and inner acceleration factors can be converted by multiplying the mass squared to lay hold of the FT gravitational time dilation factor.

Inside a sphere

At this point the equivalence of method 1 and 2 is made and the latter can be preferred given its much greater flexibility for more complex calculations. Thus by putting Equation (18) into FT's context we will have to reduce the degree of the inverse radius down to 1:

$$f = \int \frac{2\pi r_2 \left(\frac{d_2^2 - r_2^2 + r_1^2}{2d_2} + r_2 - d_2\right)}{r_2} dr_2 + \int \frac{4\pi r_2^2}{r_2} dr_2 \qquad (21)$$

$$f = \frac{4\pi d_2 (3r_1 - 2d_2)}{3} + 2\pi (r_1 - d_2)^2 \qquad (22)$$

Where:
- *r1* is the spherical mass radius
- *d2* is the distance of the clock from the center

Or more generically for a clock at a specific position inside one spherical mass, as seen from an observer positioned in a null environment:

$$t_o = -\Phi(r_s) \times t_f \qquad (23)$$

$$t_o = \frac{2\pi(3r_s^2 - r^2)}{3} \rho \times t_f \qquad (24)$$

$$t_o = \frac{m(3r_s^2 - r^2)}{2r_s^3} \times t_f \qquad (25)$$

Where:
- r is the location of the clock
- r_s is the radius of the spherical mass
- m is the mass of the sphere

For example, given a sphere of null density and radius of 30 *meters* then:

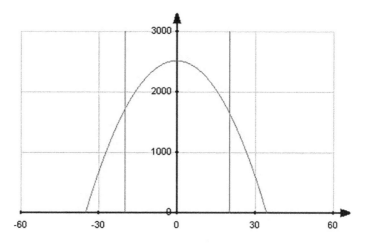

Figure 14: Inner Gravitational Time Dilation Factor vs. Radius (m)

Outside a sphere

We can now estimate the amplitude of the gravitational potential by sampling anchored bodies at an infinitesimal position by consequently rationalizing the measurement with the amplitude derived from the location of the observer.

Since an inertial body being subject to a specific gravitational force is responsible for gravitational time dilation and that gravity is a superposable force, we will translate the same conditions of all gravitational potentials into the sum of all surrounding fields of an observed clock and the observer:

$$t_o = \frac{\Phi(r)}{\Phi(r_o)} \times t_f \qquad (26)$$

$$t_o = \frac{\sum_{i=1}^{n} \frac{m_i}{|r_i - r|}}{\sum_{i=1}^{n} \frac{m_i}{|r_i - r_o|}} \times t_f \qquad (27)$$

Where:
- *r* is the location of the observed clock
- *ri* is the location of the center of mass *i*
- *ro* is the location of the observer (typically 0)
- *mi* is the mass *i*
- *to* is the observed time of two events from the clock
- *tf* is the coordinate time between two events relative to the clock

By juxtaposing the same spherical mass with an internal and external gravitational time dilation factor we have the following, for a spherical mass of 20 *meters* in radius:

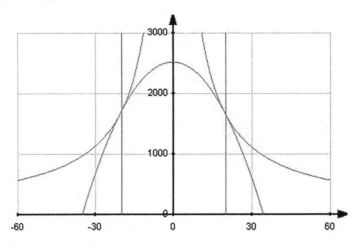

Figure 15: Inner & Outer Gravitational Time Dilation Factors vs. Radius (m)

5.4 Universe of 1 galaxy

We will now study the effects of gravitational time contraction over a bullet thrown away from the edge of a stationary galaxy. Furthermore the galaxy will be solitary in our fictional universe. Let's consider:

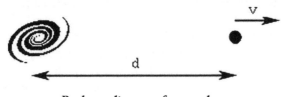

Rock traveling away from a galaxy

Where:
- m is the mass of the galaxy (1.1535736×10^{42} kg)
- r is the radius of the galaxy (4.7305×10^{20} m)
- v is the observed speed from the galaxy's edge
- x is the variable distance between the two bodies

Time contraction

To find out the gravitational time contraction factor according to the distance the bullet will be subject to we can use the simplified model, where the observer is located in a null environment for simplicity and emphasizing the tendencies:

$$t_o = \frac{m}{kx} t_f \qquad (28)$$

Where:
- $k = 1$ *kg/m*

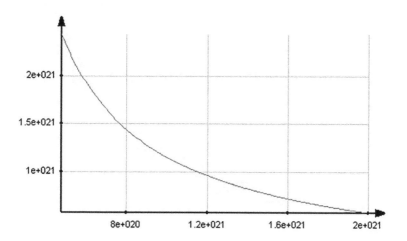

Figure 16: Time Contraction Factor vs. Distance (m)

For example this tells us a distant clock from an observer from within the galaxy's gravitational field at distance d will run faster than a local watch.

Speed contraction

Hence the speed contraction must be the exact opposite its time component because each instant is executed faster than its previous one, or are correlative:

$$v_o = \frac{t_f}{t_o} v_f \qquad (29)$$

$$v_o = \frac{kx}{m} v_f \qquad (30)$$

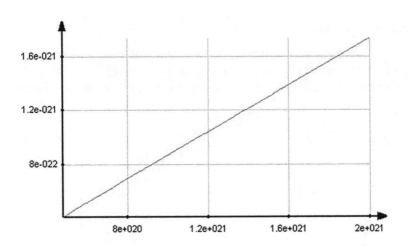

Figure 17: Speed Contraction Factor vs. Distance (m)

For example the ratio of the observed speed of the moving bullet with the speed seen 1×10^{20} m away from the outer edge of the galaxy will give us the speedup of the bullet:

$$v_f = \frac{f(4.7305\times 10^{20} m + 1\times 10^{20} m)}{f(4.7305\times 10^{20} m)} v_o \qquad (31)$$

$$v_f = 121.13941\% \times v_o \qquad (32)$$

Acceleration

We can finally confirm the constant acceleration of the bullet to be the following given our simplified model. By differentiating Equation (30) according with the position we will have an actual acceleration of:

$$a_o = \frac{k}{m} v_f \qquad (33)$$

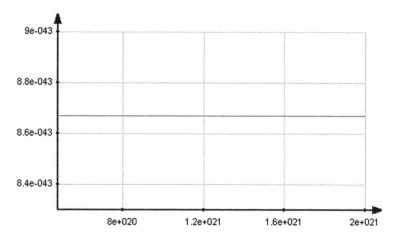

Figure 18: Acceleration Factor vs. Distance (m)

This representation is very subjective because only one galaxy is present in our schema but can symbolize nevertheless the edge of the universe if it is finite. In that case a projectile emitted from the farthest galaxy towards pure void will dramatically be accelerated in a short amount of time.

If we replace the projectile with a galaxy then the latter will have different effects, as we will shortly see.

5.5 Universe of 2 galaxies

Now we will add another galaxy that will alter the ambient gravity field and will cause the projectile to follow much different accelerated trend with less significant intensities:

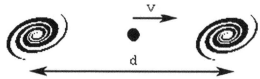

Rock traveling between galaxies

Where:
- m is the mass of the galaxy (1.1535736×10^{42} *kg*)
- r is the radius of the galaxy (4.7305×10^{20} *m*)
- v is the observed speed of the projectile
- d is the constant distance between the two galaxies (2.47305×10^{21} *m*)

General time contraction

If we sum up the amplitude of all gravitational fields, where the observer still resides a null environment then we get Figure 19.

$$t_o = \frac{t_f}{k} \times \sum_{i=1}^{n} \frac{m_i}{|x-d_i|} \qquad (34)$$

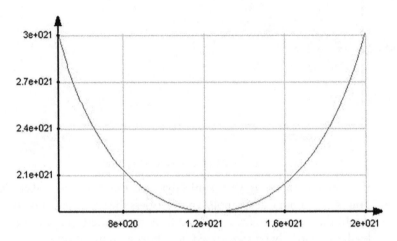

Figure 19: Mutual Time Contraction Factor vs. Distance (m)

General speed contraction

The speed will once again be inversely proportional to the time contraction. Hence Figure 20.

$$v_o = \frac{kv_f}{\sum_{i=1}^{n} \frac{m_i}{|x-d_i|}} \qquad (35)$$

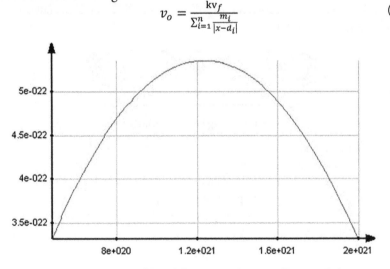

Figure 20: Mutual Speed Contraction Factor vs. Distance (m)

For example a rock thrown at 1 m/s from the edge of the leftmost galaxy towards the second one will be speeded up after it is 1×10^{20} m away by:

$$v_f = \frac{f(4.7305\times10^{20}m + 1\times10^{20}m)}{f(4.7305\times10^{20}m)} v_0 \qquad (36)$$

$$v_f = 115.08244\% \times v_0 \qquad (37)$$

General acceleration

This leads us to a very generic formulation of the accelerated universe:

$$a_o = kv_f \times \frac{\sum_{i=1}^{n} \frac{m_i}{|x-d_i|^2}}{\left[\sum_{i=1}^{n} \frac{m_i}{|x-d_i|}\right]^2} \qquad (38)$$

Which means our rock traveling between the galaxies will be subject to a slightly decreasing acceleration until midway before it finally hits the edge of the second galaxy at an accelerated speed:

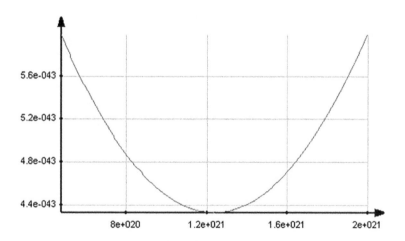

Figure 21: Mutual Acceleration Factor vs. Distance (m)

By comparing Figure 18 with 21 we can confirm by simply adding one galaxy into the representation of the general acceleration the amplitudes are much less significant than by having only one galaxy occupying the entire universe.

5.6 Dynamic universe of 2 galaxies

We were previously considering only the effects of stationary galaxies towards a moving projectile. In fact if we consider the galaxies to be moving away from each other then the gravitational field will in turn be constantly changing. This means the traveling galaxy will affect its own speed. Let's consider:

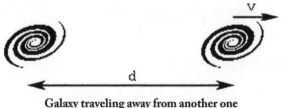

Galaxy traveling away from another one

Where:
- mi is the mass of the galaxy
- ri is the radius of the galaxy
- di is the position of the galaxy
- vf is the inertia of the moving galaxy

Dynamic speed contraction

By putting this in context and estimating the speed contraction when a galaxy itself is traveling away with constant inertia from a more massive one where gravitational forces have no effect, we will have an entire galaxy of a lesser mass that will dynamically alter the gravity field itself. Modifying the ambient gravity field according to our theory will modify the speed in regards of the galaxy in motion. Hence the moving galaxy will affect its own speed. Let's use Equation (35) to estimate the speed of a less massive moving galaxy than the leftmost one, in one instant:

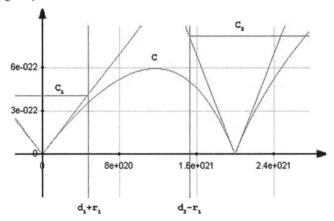

Figure 22: Dynamic Speed Contraction Factor vs. Distance (m)

Where:
- $m1 = 1.1535736 \times 10^{42}$ kg (mass of leftmost galaxy)
- $m2 = 5.767868 \times 10^{41}$ kg (mass of rightmost galaxy)
- $d1 = 0$ m (position of leftmost galaxy)
- $d2 = 2 \times 10^{21}$ m (position of leftmost galaxy)
- $r1 = r2 = 4.7305 \times 10^{20}$ m (radius of both galaxies)

- $C1$: speed contraction of leftmost galaxy
- $C2$: speed contraction of rightmost galaxy
- C: interactive speed contraction

With the above declarations we can attain the following relative speed as seen from the edge of the leftmost galaxy:

$$v_o = \frac{f(2 \times 10^{21} m - 4.7305 \times 10^{20} m)}{f(4.7305 \times 10^{20} m)} \times v_f \tag{39}$$

$$v_o = 142.61539\% \times v_f \tag{40}$$

This means the galaxy will be seen traveling 3 times faster than its local inertia from the edge of our galaxy. Now by moving the galaxy further away from us by 2.5×10^{20} m we will quickly see the different apparent speed it will carry:

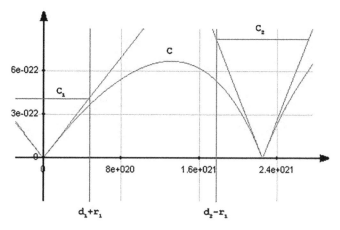

Figure 23: Dynamic Speed Contraction Factor vs. Distance (m)

$$v_o = \frac{f(2.25 \times 10^{21} m - 4.7305 \times 10^{20} m)}{f(4.7305 \times 10^{20} m)} \times v_f \tag{41}$$

$$v_o = 147.88381\% \times v_f \tag{42}$$

We already see here an acceleration in the perceived speed of the moving galaxy by 5.26842% for a distance of 2.5×10^{20} m away from its original position[1]. To better estimate the perceived speed of the galaxy according to a variable position we will use the consequent formula where the speed sample taken from the gravity field will always be on the innermost radius of the galaxy. What changes here is the position of the traveling galaxy, which is variable:

$$v_o = \frac{kv_f}{\frac{m_1}{|x-d_1|} + \frac{m_2}{|r_2|}} \qquad (43)$$

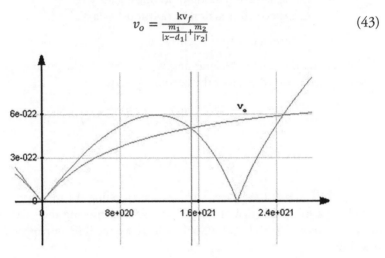

Figure 24: Dynamic Speed Contraction Factor vs. Distance (m)

Irrevocably the speed of the nomadic galaxy with an initial inertia will be greatly enhanced the farther it gets away from us. There is absolutely no repulsive force necessary to accomplish this behavior.

Furthermore this concept will obviously to even greater scales such as cluster of galaxies, superclusters and even greater probable groups of superclusters. The discussion on knowing how large the universe is not conclusive enough unless possible reverse engineering is used to map the measurements to approximate the direction of the galaxies and therefore the center of the universe if a big bang is really responsible for its creation.

Dynamic acceleration

In our estimates we are taking into account only 2 galaxies. Once again this drift can be applied to greater scales in more important amplitudes but they will still follow the same course. This course can be estimated as such:

$$a_o = kv_f \times \frac{m_1}{\left[\frac{m_1}{|x-d_1|} + \frac{m_2}{|r_2|}\right]^2 (x-d)^2} \qquad (44)$$

---

1   This represents a universe of 2 galaxies, not the actual universe of 7 trillion galaxies!

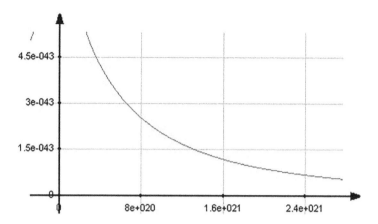

Figure 25: Dynamic Acceleration Factor vs. Distance (m)

This graph summarizes how gigantic masses will interact with each other. The acceleration is not constant in short distances in contrast with the Hubble's Law but will eventually tend to be this way for very long distances.

Moreover the lighter the traveling galaxy is the faster it will speed up in relation to a more massive cluster or supercluster. This means dwarf galaxies are subject of being much distant than massive ones if the big bang theory announce itself to be true.

Finally if the universe is finite and has a edge then all tangential galaxies will be ejected even faster than innermost ones because of the average gravitational field being more important towards the center of the universe. Section 5.4.2 still applies in these conditions but at the scale of the entire universe.

5.7 Time dilation effect

Kinematical time dilation

We can represent time dilation using simpler techniques by interpolating dilation. Indeed if we rationalize the kinetic energy gained by the object in motion according to the maximum one it can experience at the speed of light then, due to the Hypothesis 2, we have

$$p_v = \frac{mv^2/2}{mc^2/2} \qquad (45)$$

Since the time dilation percentage is the exact opposite of the speed ratio, we define general time dilation in direct relation to the proportion as follows:

$$\frac{\Delta\tau_v}{\Delta\tau_0} = 1 - p_v = 1 - \frac{v^2}{c^2} \qquad (46)$$

Here, $\Delta\tau_v$ is the interval of time between some events measured in the proper reference of moving observer and $\Delta\tau_0$ is interval of time between the same events measured by the static observer. is the relative velocity of moving observer measured by the static one and $c = 2.998 \times 10^8$ m/s is the speed of light.

We can note that the Finite Theory prediction (46) contradicts to the special relativistic result

$$\frac{\Delta\tau_v}{\Delta\tau_0} = \sqrt{1 - \frac{v^2}{c^2}} \approx 1 - \frac{v^2}{2c^2}, \qquad (47)$$

where the last equality is valid for small velocities. Nevertheless, as we will see in Sec. 5.10.2, when the kinematical time dilation effect is combined with the gravitational one, Finite Theory predicts absolutely correct value of the time dilation cancellation altitude, which is observed by GPS satellites. In the following, we will investigate the gravitational time dilation effect in more detail.

Gravitational time dilation

Effect of the time dilation in the gravitational field is a consequence of the difference in gravitational potentials. This effect is described by the relation

$$\frac{\Delta\tau}{\Delta t} = \frac{1}{h}\left(h + \frac{M}{r}\right) = 1 + \frac{M}{hr} \qquad (48)$$

where, is a mass of the gravitating object and is the distance from its centre. Under $\Delta\tau$ we mean the interval of local time at the point situated at distance r from the centre of the source of gravitation. Δt is the interval of time measured by the distant observer, situated at distance $r \to \infty$.

General relativistic time dilation effect is a particular case of (48) if $h=c^2/G$, where is the speed of light and $G = 6.674 \times 10^{(-11)}$ m^3 kg$^{(-1)}$ s$^{(-2)}$ is the gravitational constant. Indeed, we know that in the weak field limit of General Relativity, time dilation effect in the gravitational field takes the following form (see, for example, [7]):

$$\frac{\Delta\tau}{\Delta t} = \left(1 + \frac{GM}{c^2 r}\right)^{-1} \approx 1 - \frac{GM}{c^2 r} \qquad (49)$$

But due to the hypotheses of the Finite Theory, factor in (48) is not a universal constant but depend on the superposed gravitational potentials. For example, in solar system experiments, where the gravitational potential of the Sun is the source of the strongest gravitational acceleration, we suppose. The value of can be determined from the observation of the deflection angle of light in the gravitational field of the Sun, as we will demonstrate in the next subsection.

5.8 Bending of light in the gravitational field

Due to the time dilation effect, we expect to have different speed measurements of the same body by different observers. In particular, the speed of light traveling through the gravitational field will be different from the viewpoint of a local observer and from the viewpoint of a distant watcher.

According to (48), a distant observer notes that the light beam has a velocity, which depends on the position in the gravitational field:

$$v = \frac{dr}{dt} = \frac{dr}{d\tau}\left(1 + \frac{m_{sun}}{h_{solar}r}\right) = c\left(1 + \frac{m_{sun}}{h_{solar}r}\right) \quad (50)$$

In this relation, the local speed $v_{local} = dr/d\tau = c = 2.998 \times 10^8$ m/s is constant due to our hypothesis (Hypothesis 1). Also, we neglect the effect of length contraction in the gravitational field, which results in the equal values of length interval dr for both local and distant observers.

Distant observer can interpret the slow down of the light speed as the effect of some nonzero effective index of refraction:

$$n(r) \equiv \frac{c}{v} = \left(1 + \frac{m_{sun}}{h_{solar}r}\right)^{-1} \approx \left(1 - \frac{m_{sun}}{h_{solar}r}\right) \quad (51)$$

The last approximate relation here is due to the fact we suppose $|m_{sun}/h_{solar}| \ll r$. As we will see later, this condition is fulfilled for the majority of real astrophysical objects.

The position dependent index of refraction causes the bending of light, which will be measured by distant observer. For the refractive index (51), the value of deflection angle is as follows:

$$\delta = \frac{4m_{sun}}{h_{solar}r_{sun}}, \quad (52)$$

where is the impact parameter, or the minimal distance from the light ray to the source of gravity. This relation is an obvious generalization of the result derived by Einstein. More detailed derivation can be found in [8].

Observed value of the deflection angle equals to (see [9], [10])

$$\delta_{obs} = \frac{4Gm_{sun}}{c^2 r_{sun}} = 0.847 \times 10^{-5} \text{ rad} \quad (53)$$

Both General Relativity and Finite Theory can adjust the theoretical result (54) with the observed value (53), but in different ways:

1. To explain the experiment in General Relativity, which supposes $h_{solar} = c^2/G = 1.35 \times 10^{27}$ kg/m, we have to introduce additional length contractions in the gravitational field, as is explained in [7].
2. In Finite Theory, we are decomposing the deflection angle into the time dilation and Newtonian acceleration constituents:

$$\delta = \frac{2m_{sun}}{h_{solar}r_{sun}} + \frac{2m_{sun}}{h_{solar}r_{sun}} = \frac{4m_{sun}}{h_{solar}r_{sun}}, \qquad (54)$$

No additional length contractions in the gravitational field is required in this case.

5.9 Explanation of the perihelion shift

The bending of light and perihelion shift of planets are the two classical tests of General Relativity. As we have seen in the previous subsection, bending of light can be naturally explained by the Finite Theory without length contractions in the gravitational field. In this section, we consider the possibility for the Finite Theory to explain the perihelion shift of planets.

As we know, the radial motion of a planets in the gravitational field of the Sun in Newton's gravity can be described by the relation

$$\frac{m\dot{r}^2}{2} + V(r) = \mathcal{E}, \qquad (55)$$

where $V(r)$ is defined by

$$V(r) = -\frac{GmM}{r} + \frac{L^2}{2mr^2} \qquad (56)$$

Here, m is a mass of planet, M — mass of the Sun, $\mathcal{E}$ — full non-relativistic energy of the planet, and L is the value of conserved angular momentum. Variable $r=|\vec{r}|$ is the distance to the Sun, which is supposed to be situated in the centre of coordinate system, and the dot means differentiation with respect to . The second term in , in contrast to the attractive Newton's potential (first term), describes the action of repulsive centrifugal forces.

The general-relativistic investigation of the trajectory of a massive object in the spherically-symmetric gravitational field can also be described in terms of the effective gravitational potential (see, for example, [7]):

$$\frac{m\dot{r}^2}{2} + V_{eff}(r) = \mathcal{E}, \qquad (57)$$

$$V_{eff}(r) = -\frac{GmM}{r} + \frac{L^2}{2mr^2}\left(1 - \frac{2GM}{c^2r}\right) \qquad (58)$$

Thus, the effective gravitational potential of General Relativity using the gravitational time dilation substitution (7), can be written in the form

$$V_{eff}(r) = -\frac{GmM}{r} + \frac{L^2}{2mr^2}\left(1 - \frac{2GM}{c^2 r}\right) \quad (59)$$

$$V_{eff}(r) = -\frac{GmM}{r} + \frac{L^2}{2mr^2}\left(1 + \frac{M}{h_{solar}r}\right)^{-2} \quad (60)$$

As is demonstrated in [7], such correction to the gravitational potential leads to the perihelion shift of the elliptical orbit per unit revolution by the angle

$$\delta\varphi = \frac{6\pi GM}{c^2 a(1-e^2)}, \quad (61)$$

where a is the semi-major axis of the orbit and e is it's eccentricity.

We know (see [9], [10]) that the perihelion shift agrees with observational evidences not only for the Mercury, but for all planets of solar system. Thus, the perihelion shift can be successfully explained within a Newtonian mechanics if the correction (60) to the Newtonian potential energy is taken into account. This work has demonstrated that the additional term in (60) can appear as the result of the velocity-dependent correction that acts on planets in solar system.

5.10 GPS and time dilation cancellation altitude

The gravitational time dilation and the kinematical time dilation both play a role on GPS satellites. The former is affected by the altitude whereas the latter is affected by its speed. We will study here the correct altitude where both effects cancel out.

First, we consider time dilation cancellation altitude from the viewpoint of General Relativity.

Time dilation cancellation altitude in General Relativity

Consider the artificial satellite, rotating around the Earth on circular orbit with radius r_{orbit}. Due to the gravitational dilation of time [see (49)], static observer at altitude $r_{orbit} > r_{earth}$ should feel accelerated flow of time with respect to the static observer on the Earth (r_{earth} is the radius of the Earth):

$$\frac{\Delta\tau_{orbit}}{\Delta\tau_{earth}} = \sqrt{\frac{1 - \frac{2Gm_{earth}}{c^2 r_{orbit}}}{1 - \frac{2Gm_{earth}}{c^2 r_{earth}}}} \quad (62)$$

But satellite is not static, it rotates with linear velocity v, which leads to additional relativistic effect:

$$\frac{\Delta\tau_v}{\Delta\tau_{earth}} = \sqrt{1 - \frac{v^2}{c^2}} \approx 1 - \frac{v^2}{2c^2} \qquad (63)$$

Here, we are using the low-velocity approximation ($v \ll c$), which is justified for real GPS satellites. As we can see, relativistic effect is opposed to the gravitational one, which makes it possible to find altitude, at which time dilation is cancelled.

Finale relation, which takes into account both effects, can be written in the form:

$$\frac{\Delta\tau_{satellite}}{\Delta\tau_{earth}} = \sqrt{\frac{1 - \frac{2Gm_{earth}}{c^2 r_{orbit}}}{1 - \frac{2Gm_{earth}}{c^2 r_{earth}}}} \left(1 - \frac{v^2}{2c^2}\right) \approx 1 + \frac{Gm_{earth}}{c^2 r_{orbit}} - \frac{Gm_{earth}}{c^2 r_{earth}} - \frac{v^2}{2c^2} \qquad (64)$$

where the last approximate equality is valid in the Newtonian limit $r_{earth}, r_{orbit} \gg m_{earth}/h$. Also, under these conditions we can use the Newtonian relation for the velocity of satellite, rotating on the circular orbit $v^2 = Gm_{earth}/r_{orbit}$, which results in the relation

$$\frac{v^2}{c^2} = \frac{Gm_{earth}}{c^2 r_{orbit}} \qquad (65)$$

Consequently, the radius of orbit, at which cancellation occurs, is found to be

$$\frac{\Delta\tau_{satellite}}{\Delta\tau_{earth}} \approx 1 - \frac{3Gm_{earth}}{2c^2 r_{orbit}} + \frac{Gm_{earth}}{c^2 r_{earth}} = 1 \;\Rightarrow\; r_{orbit} = \frac{3r_{earth}}{2} \qquad (66)$$

which corresponds to the altitude $H = r_{orbit} - r_{earth} = r_{earth}/2 \approx 3185$ km [1].

Time dilation cancellation altitude in Finite Theory

For the same artificial satellite, Finite Theory supposes the gravitational dilation of time for static observers to be defined by [see (48) and (8)]

$$\frac{\Delta\tau_{orbit}}{\Delta\tau_{earth}} = \frac{1 + \frac{m_{earth}}{h_{solar} r_{earth}}}{1 + \frac{m_{earth}}{h_{solar} r_{orbit}}}, \quad h_{solar} = \frac{c^2}{G} \qquad (67)$$

You'll notice above that the radiuses are swapped when compared to its GR counterpart (5). For the kinematical time dilation effect in Finite Theory we have (see the explanation in Sec. 5.7.1):

$$\frac{\Delta\tau_v}{\Delta\tau_{earth}} = 1 - \frac{v^2}{c^2} \qquad (68)$$

Though both kinematical and gravitational time dilation effects predicted by Finite Theory differ from those effects in General Relativity, combined effect to the artificial satellite appears to be the same in both theories. Indeed, combining (67) and (68) we get

$$\frac{\Delta\tau_{satellite}}{\Delta\tau_{earth}} = \frac{\left(1+\frac{m_{earth}}{h_{solar}r_{earth}}\right)\left(1-\frac{v^2}{c^2}\right)}{1+\frac{m_{earth}}{h_{solar}r_{orbit}}} \quad (69)$$

For the orbital velocity of satellite we have $v^2 = Gm_{earth}/r_{orbit}$, which results in the relation

$$\frac{v^2}{c^2} = \frac{Gm_{earth}}{c^2 r_{orbit}} = \frac{m_{earth}}{h_{solar}r_{orbit}} \quad (70)$$

Thus, we can write

$$r_{orbit} = \frac{2h_{solar}r_{earth} + m_{earth}}{h_{solar}} \quad (71)$$

Cancellation effect take place at altitudes where $\Delta\tau_{satellite} = \Delta\tau_{earth}$. Corresponding altitude $H = r_{orbit} - r_{earth} = 6371$ km absolutely coincides with twice the altitude derived in Sec. 5.10.1 in the frames of General Relativity. In other words, this prediction can be upgraded into yet another experiment proposal.

5.11 Gravitational redshift

As we know, the relativistic Doppler shift is given by the following formula:

$$f_o = \sqrt{\frac{1-\frac{v}{c}}{\frac{v}{c}+1}} \times f_f \quad (72)$$

In contrast with the Finite Theory's kinetic Doppler shift as given by:

$$f_o = \frac{1}{\frac{v}{c}+1} \times f_f \quad (73)$$

Which will result in the following divergent functions:

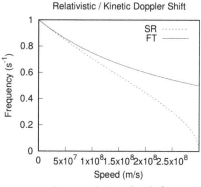

Figure 26: Doppler Shift

Thus at low velocities both functions are equivalent but at high velocities they diverge. Unfortunately testing the gravitational redshift at high velocities is still a problem as of today so we'll have to restrict ourselves to testing low velocities.

In order to find the gravitational redshift and the relativistic Doppler shift cancelation point, we'll use the following formula:

$$\sqrt{\frac{\left(1-\frac{2G\,m_e}{c^2\,|p_e-d|}\right)\left(1-\frac{v}{c}\right)}{\left(1-\frac{2G\,m_e}{c^2\,|p_e|}\right)\left(\frac{v}{c}+1\right)}} = 1 \qquad (74)$$

Where:
- v is the relative velocity between the observer and the moving apparatus
- $c = 299792458$ m/s
- $G = 6.67408$ m^3 kg^{-1} s^{-2}
- $d = 22.5$ m (elevation of the tower)
- $p_e = -6.371 \times 10^6$ m (position of the center of the Earth)
- $m_e = 5.973 \times 10^{24}$ kg (mass of the Earth)

Or:

$$v = \frac{G c d\, m_e}{c^2\,|p_e|^2 + (c^2 d - 2G\,m_e)|p_e| - G d\, m_e} \qquad (75)$$

$$v = 7.322 \times 10^{-7} \text{ m/s} \qquad (76)$$

$$v/c = 2.442 \times 10^{-15} \qquad (77)$$

In terms of Finite Theory's gravitational redshift and the kinetic Doppler shift cancelation point, we'll use in turn the following formula:

$$\frac{\left(\frac{m_e}{|p_e|} + h_{solar}\right)\left(1-\frac{v}{c}\right)}{\left(\frac{m_e}{|p_e-d|} + h_{solar}\right)\left(1-\frac{v^2}{c^2}\right)} = 1 \qquad (78)$$

Where:
- v is the relative velocity between the observer and the moving apparatus
- $c = 299792458$ m/s
- $G = 6.67408$ m^3 kg^{-1} s^{-2}
- $h_{solar} = c^2/G = 1.35 \times 10^{27}$ kg/m
- $d = 22.5$ m (elevation of the tower)
- $p_e = -6.371 \times 10^6$ m (position of the center of the Earth)
- $m_e = 5.973 \times 10^{24}$ kg (mass of the Earth)

Or:

$$v = \frac{c\, m_e |p_e - d| - c\, m_e |p_e|}{h_{solar} |p_e||p_e - d| + m_e |p_e|} \qquad (79)$$

$$v = 7.322 \times 10^{-7}\ m/s \qquad (80)$$

$$v/c = 2.442 \times 10^{-15} \qquad (81)$$

We'll notice once again that by using completely different mathematics we obtain exactly the same factors which correspond to observations [5].

– 6 –
Cosmological implications

Herein are enumerated all consequences FT will lead to and highlights important differences from GR. Given we know the measurement of the light bending, we can "reverse engineer" the entire universe to find out all its characteristics. We'll now illustrate how it can be done.

At this level only complex computer research can be proposed to simulate a modeling of the universe under this umbrella in order to match its behavior with measurements such as the constant of Hubble's Law. Potentially, simulators can also be used to reverse time and estimate an early universe according to the current velocities of the superclusters, solve the scaling factor of the observed universe which will lead to an estimation of the real volume of the universe and solve local focal points of gravitational lenses.

6.1 Natural faster-than-light speed

One of the most practical and interesting goals of any research area in this field is to reach exoplanets. Unfortunately since GR disallows any probe or ship to travel faster than we reach an impasse because one of the closest star named Alpha Centauri is about **4.36507** *ly* or **4.01345** *m* away from us. This means light rays will take **4.36507** *years* to overtake that distance according to GR. The following section explores the consequences of FT on close distances such as the Moon that will follow the following principle:

$$t = \int \frac{\sum_{i=1}^{n} \frac{m_i}{|x - d_i|}}{\sum_{i=1}^{n} \frac{m_i}{|d_i|}} \times \frac{1}{c} dx \qquad (82)$$

Moon

In order to estimate the distance of the Moon in conformance to FT, we will follow the henceforth equation that takes into account the adjoining most massive entity, or the influence of the scaling factor. We also know the time it takes for a laser to travel back and forth between the Moon and the surface of the Earth. Once again the scaling factor represents the average influence of all surrounding stars:

$$1.25\ s = \frac{m_s \log(|x_{ft}-r_m-p_s|)+m_e\log(|x_{ft}-r_m-p_e|)}{c\left(\frac{m_m}{|x_{ft}|}+\frac{m_s}{|p_s|}+\frac{m_e}{|p_e|}+h_{solar}\right)}$$
$$+ \frac{h_{solar}|x_{ft}-r_m|+m_m\log(|r_m|)}{c\left(\frac{m_m}{|x_{ft}|}+\frac{m_s}{|p_s|}+\frac{m_e}{|p_e|}+h_{solar}\right)}$$
$$- \frac{m_m\log(|x_{ft}|)+m_s\log(|p_s|)+m_e\log(|p_e|)}{c\left(\frac{m_m}{|x_{ft}|}+\frac{m_s}{|p_s|}+\frac{m_e}{|p_e|}+h_{solar}\right)} \qquad (83)$$

And after numerical analysis we'll find that:

$$x_{ft} = 3.7647807986 \times 10^8\ m \qquad (84)$$

Where:
- $c = 299792458$ m/s
- $G = 6.67408$ $m^3\ kg^{-1}\ s^{-2}$
- $h_{solar} = c^2/G = 1.35 \times 10^{27}$ kg/m
- $p_e = -6.371 \times 10^6$ m (position of the center of the Earth)
- $m_e = 5.973 \times 10^{24}$ kg (mass of the Earth)
- $r_m = 1.7375 \times 10^6$ m (radius of the Moon)
- $m_m = 7.348 \times 10^{22}$ kg (mass of the Moon)
- $p_s = 1.52 \times 10^{11}$ m (position of the center of the Sun)
- $m_s = 1.98892 \times 10^{30}$ kg (mass of the Sun)

If we compare with the distance predicted by GR:

$$x_{gr} = c \times 1.25 + r_m \qquad (85)$$

$$x_{gr} = 3.764780725 \times 10^8\ m \qquad (86)$$

Which is a difference of:

$$x_{ft} - x_{gr} = 7.36\ m \qquad (87)$$

Indeed we just found a discrepency of between the prediction of FT and GR at such a low scale.

6.2 Parameters of the invisible universe

Fudge factor of the invisible universe

An inside-the-sphere gravitational potential distribution formula of the entire visible universe predicts the following value of the parameter h:

$$h_{visible} = \frac{M_{visible}(3R_{visible}^2 - d^2)}{2R_{visible}^3} \tag{88}$$

Here, $M_{visible} = 10^{53}$ kg is the mass of the entire visible universe, $R_{visible} = 4.4 \times 10^{26}$ m is it's radius, and d is the location of the Milky Way in the visible universe. In the following, we suppose $d = 0$ m. Thus, we can calculate

$$h_{visible} = \frac{3M_{visible}}{2R_{visible}} = 0.34 \times 10^{27} \, kg/m \tag{89}$$

As we can see, the value of $h_{visible}$ does not coincides to the value $h_{solar} = 1.35 \times 10^{27}$ kg/m which was derived in Sec. 0.5.8. This can be explained by the presence of some invisible constituents of the universe. If so, we can decompose

$$h_{solar} = h_{visible} + h_{invisible} \tag{90}$$

Thus by solving $h_{invisible}$ we obtain

$$h_{invisible} = h_{solar} - h_{visible} = 1.01 \times 10^{27} \, kg/m \tag{91}$$

In the following we will use the obtained value of $h_{invisible}$ to determine the mass of the invisible universe.

Mass of the invisible universe

Since the invisible universe will follow the same inside-a-sphere distribution as the visible one, then

$$h_{invisible} = \frac{3M_{invisible}}{2R_{invisible}}, \tag{92}$$

which results in

$$M_{invisible} = \frac{2h_{invisible}R_{invisible}}{3} = 7.38 \times 10^{55} \, kg \tag{93}$$

That means the mass of the invisible universe is times the mass of the visible universe. To calculate the value of $M_{invisible}$, we have supposed $R_{invisible} = 1.1 \times 10^{29}$ m and used the result obtained in (91).

To compute $M_{invisible}$ directly from the light bending δ we can also use the following relation:

$$M_{invisible} = \frac{R_{invisible}(4R_{visible}m_{sun} - 3\delta r_{sun}M_{visible})}{3\delta r_{sun}R_{visible}} \quad (94)$$

6.3 Galactic rotation curve

Altough we've been taking for granted classical physics was flawless we need to revert the last centuries of research in physics in order to admit the mistakes in the bedrock of physics. Indeed, the laws of Newton imply there is a parent framework from which rotating stars relates to in all cases. But what if we have a universe with only one black hole in it? The black hole will rotate relative to which comoving framework?

We can clearly see here that the black hole cannot rotate if it is the sole one in the universe. As a matter of fact, this singular black hole will be the one defining the absolute framework from which other stars will rotate around and their planets will rotate around each one of these stars, and so on. So the comoving frameworks simply are superposed with the most massive body defining the absolute one.

The best example we can use to prove this property is indeed the rotation curve because of the huge mass involved at the center of galaxies. If we take for example the formula of a standard rotation curve as given by the following [4]:

$$y_1 = \sqrt{\frac{i \, m_f m_{star}}{h \sqrt{\tan\left(\frac{\pi e_{max} i}{2 n_p}\right)}}} \quad (95)$$

$$i = \frac{2 n_p atan\left(\frac{x^2}{h^2}\right)}{\pi e_{max}} \quad (96)$$

$$e_{max} = \frac{2 atan\left(\frac{r_{max}^2}{h^2}\right)}{\pi} \quad (97)$$

$$m_t = \frac{4 r_{max} atan\left(\frac{r_{max}}{r_0}\right)^2 v_0^2}{\pi^2} \quad (98)$$

$$m_{star} = \frac{m_t}{8 n_p} \quad (99)$$

Here n_p = 200 is the number of stars in the galaxy, r_{max} = 20 kpc is the maximum radius of the galaxy, h = 4.5 kpc is the radius of the bulge, v_0 = 140 km/s is the tangential velocity of the closest star and r_0 = 1 kpc is the position of the closest star. This will result in Figure 27.

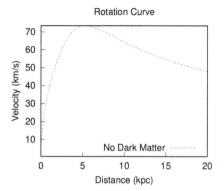

Figure 27: Rotation Curve

Now let's define the dark matter contribution with:

$$y_2 = \sqrt{\frac{f_{dm}(m_t - f_m m_{star} n_p)\left(\frac{x}{r_{dmo}} - atan\left(\frac{x}{r_{dmo}}\right)\right)}{\left(\frac{r_{max}}{r_{dmo}} - atan\left(\frac{r_{max}}{r_{dmo}}\right)\right)x}} \qquad (100)$$

Here $f_{dm} = 0.1$ is the dark matter mass scale, $f_m = 1$ is the matter mass scale and $r_{dmo} = 1\ kpc$ is the dark matter length scale.

By adding the dark matter contribution to the visible matter then we will have:

$$y_3 = \sqrt{y_2^2 + y_1^2} \qquad (101)$$

Which will result in Figure 28.

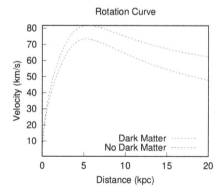

Figure 28: Rotation Curve

Now to compute the Finite Theory rotation curve, all we have to do is to add a spin to the entire frame of reference:

$$y_4 = y_1 + \omega_{FT} x \qquad (102)$$

Here $\omega_{FT} = 1.5 \text{ s}^{-1}$ is added angular velocity to the frame of reference. This will result in Figure 29.

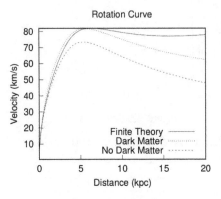

Figure 29: Rotation Curve

The aforementioned graphic clearly demonstrates the veracity of Finite Theory in this scenario as well. In this demonstration there was some arbitrary factors being used but the general idea being proved remains the same.

6.4 Approximation of the center and velocity of the visible universe

Dark energy is a constant or scalar field filling all of space that has been hypothesized but remains undetected in laboratories. The problem is that in order to do so the amount of vacuum energy required to overcome gravitational attraction would require a constantly increasing total energy of the universe in violation of the law of conservation.

Small scales

The Hubble's law represents the rate of the expansion of the universe with the speed of the distant galaxies $v_{apparent}$ as seen from the Milky Way with:

$$v_{apparent} = H_0 \, x , \qquad (103)$$

where $H_0 = 2.26 \times 10^{-18}$ s is a Hubble's constant and x is a distance to the remote galaxy. Hubble law is illustrated in Fig. 30.

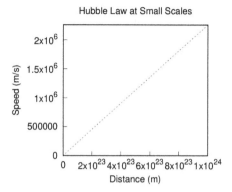

Figure 30: Hubble law

On the other hand Finite Theory applied on the scale of the universe proves that there is no need for such energy. Indeed if we consider the universe to be the result of a big bang then all galaxies must have a certain momentum. If we try to represent the speed of the observed galaxies using Finite Theory where h is null because the environment must not be encompassed by anything else then we will have:

$$v_{apparent} = \frac{M_{visible}/|s_{visible}|}{M_{visible}/|x-s_{visible}|} v_{visible} \quad (104)$$

where $s_{visible}$ is a position of the center of the visible universe, and $v_{visible} = c$.

After simplifying and subtracting the speed of the observer from his observations[2] we will have:

$$v_{apparent} = \frac{v_{visible}|x-s_{visible}|}{|s_{visible}|} - v_{visible} \quad (105)$$

This means $s_{visible}$, or the position of the center of the universe, is actually solvable by equaling (103) and (105):

$$H_0 x = \frac{v_{visible}|x-s_{visible}|}{|s_{visible}|} - v_{visible}, \quad (106)$$

which results in

$$s_{visible} = -\frac{v_{visible}}{H_0} \quad (107)$$

$$s_{visible} = -1.33 \times 10^{26} \, m \quad (108)$$

---

[2] The speed of the observer $v_{visible}$ needs to be subtracted because the observer himself is moving and has the same speed of the visible universe ($v_{visible}$).

Obviously the aforementioned speed is unidimensional and therefore has no velocity vector meaning there is no way to tell its direction given the isotropism of the visible universe.

- 7 -
Experiment proposal

Although gravitons have not been directly detected and might not even be possible [6], we hypothesize to detect its presence indirectly by observing a variance in both c and the wavelength of a photon from the graviton field it is traveling through. We reevaluate the absoluteness of the reference frames, as is demanded by the hypotheses of the Finite Theory.

Since gravity obeys the principle of superposition, we will have to isolate which reference frame defines the absoluteness of the kinetic time dilation amplitude via the gravitational acceleration strength:

$$a_{earth} = - \frac{Gm_{earth}}{(x-p_e)^2}, \qquad (109)$$

$$a_{sun} = - \frac{Gm_{sun}}{(x-p_s)^2}, \qquad (110)$$

Here, is the mass of the Earth, is the mass of the Sun, is a position of the center of the Earth and is a position of the Sun. The behavior of both accelerations is illustrated in Fig. 31.

Thus the reference frame for altitudes lower than the following is defined by the Earth:

$$x = \frac{(p_s-p_e)\sqrt{m_{earth} \times m_{sun}}+p_e \times m_{sun}-p_s \times m_{earth}}{m_{sun}-m_{earth}} = 2.5245 \times 10^8 \, m \qquad (111)$$

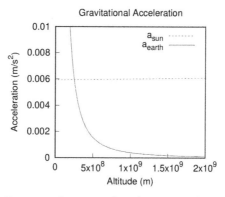

Figure 31: Gravitational acceleration vs. Altitude

The observer is subject to time dilation relative to the surface of the Earth but the wavelength meter is also subject to the exact same amount of time dilation so both effect cancel out and what the observer sees is a normally functioning wavelength meter. The wavelength is relative to the spinning surface of the Earth so having an observer moving against it will change what is measured. Also the frequency (cycles per second) will be the same in all frames of reference. The hypothesis related to time dilation have no effect here and only the hypothesis related to the frames of reference play a role.

By sending the experiment at a speed in the vicinity of the speed of sound (in the following, we suppose the speed of the experimenter to be), it should be sufficient to detect a change wavelength directly proportionally while energy is conserved:

$$E = \frac{h(c-v_1)}{\lambda_1} = \frac{h(c-v_2)}{\lambda_2} \quad (112)$$

Thus, if the stationary observer () measures , experimenter having velocity measures

$$\lambda_2 = \frac{(c-v_2)\lambda_1}{c-v_1} = 6.49987 \times 10^{-7} \, m \quad (113)$$

Here, we have accepted for the local value of the light speed.

As the frequency will be the same in all frames of reference, the speed of light won't be constant, relative to the moving observer. For the stationary observer, which measures speed of light and wavelength , we have

$$v_1 = \frac{c_1}{\lambda_1} = 4.6122 \times 10^{14} \, s^{-1} \quad (114)$$

Now we can find the new speed of the light beam in motion, which will be measured by an experimenter having velocity :

$$c_2 = \lambda_2 v_2 = \lambda_2 v_1 = 2.9979 \times 10^8 \, m/s, \quad (115)$$

where we have combined results (113) and (114).

For a wavelength meter having an accuracy of we should be able to confirm whether the change in wavelength (and, correspondingly, the change of light speed) occurs for the experiment in motion. The predicted difference of is large enough to be detected [3].

- 8 -
Conclusion

As we have demonstrated in this proposal, Finite Theory is a viable candidate to the new theory of gravity, which can explain time dilation effects, bending of light and perihelion shifts for planets in solar system (see Sec. 5). Also, Finite Theory allows to establish new properties of the invisible part of the universe and explain some peculiar properties of late-time cosmological evolution (Sec. 6).

Though we still have some unresolved problems, we believe that results obtained to this moment are very promising, and Finite Theory deserves for further theoretical and experimental investigation. The role of the experiment we have described in Sec. 7 is crucial for the development of the Finite Theory. Possibly, it will start new era in the gravitational physics.

References

[1] N. Ashby. Relativity and the global positioning system, 2002.

[2] D. Y. Gezari. *Experimental Basis for Special Relativity in the Photon Sector.* [physics.gen-ph], 2009.

[3] HighFinesse. Highfinesse wavelength meters. 2019.

[4] A. Pisani. Galaxy rotation with dark matter. Simulator, 2014.

[5] Robert V. Pound and J. L. Snider. Effect of Gravity on Nuclear Resonance. *Phys. Rev. Lett.*, 13:539–540, 1964.

[6] T. Rothman and S. Boughn. Can gravitons be detected? *Foundations of Physics*, 2006.

[7] Ta-Pei. *Relativity, Gravitation and Cosmology. A Basic Introduction.* Oxford University Press, 2005.

[8] Ta-Pei. *Einstein's Physics. Atoms, Quanta, and Relativity Derived, Explained, and Appraised.* Oxford University Press, 2013.

[9] C. M. Will. *Theory and Experiment in Gravitational Physics.* Cambridge University Press, 1993.

[10] C. M. Will. The confrontation between general relativity and experiment. *Living Rev*, 17, 2014.

Index

A

Acceleration 16
  Dynamic 22
  General 19
Acceleration factor
  Inner 9
  Juxtaposition 11
  Outer 12
Ambient gravity field 8, 17, 20

B

Big bang 22, 23, 39
Black hole
  Radius 8

C

Cannon 3, 4

F

Finite Theory 4, 7, 24, 25, 26, 28, 29, 30, 37, 38, 39, 41, 43
Fudge factor 35

G

Galaxy 14, 20
  Dwarf 23
  Ejected 23
General Relativity 4, 5, 7, 24, 25, 26, 27
  Comparison 4
Gravitational constant 24
Gravitational fields
  Amplitude 41
Gravitational time contraction 8
  Outside a sphere 13

H

Hubble 23
Hubble's Law 23, 33

I

Implications
  Finite Theory 37

L

Lorentz transformations 1

N

Newton
  Gravitational acceleration 11

S

Schwarzschild 8
Spaceship 2
Special Relativity 2
  Fails 2
  Paradox 2
Speed contraction 15
  Dynamic 20
  General 18
Speed of light ix, 1, 2, 3, 7, 8, 23, 24

T

Thought experiment 3, 4
Time contraction 15
  General 18
Time dilation 23
  Lorentz 1
Twin paradox 2

U

Universe 19, 22, 23
　1 galaxy 14
　2 dynamic galaxies 20
　2 galaxies 17
　Center 22
　Edge 14
　Modeling 33

CPSIA information can be obtained
at www.ICGtesting.com
Printed in the USA
LVHW041547130820
663094LV00004B/681